£5-00

SBN 900848 07 3
PRINTED IN ENGLAND AT
THE CURWEN PRESS, PLAISTOW, E13
ON PAPERS MADE BY GROSVENOR CHATER
THIS EDITION IS LIMITED TO 1000 COPIES
OF WHICH THIS IS NUMBER

CLASSICA ENTOMOLOGICA

A SHORT

HISTORY

OF THE

BROWN-TAIL MOTH

[1782]

By WILLIAM CURTIS

AUTHOR OF THE FLORA LONDINENSIS

WITH AN INTRODUCTION BY

WILLIAM T. STEARN

AND ENTOMOLOGICAL NOTES BY

D. S. FLETCHER

Illustrated by a COLOUR PLATE

CURWEN FACSIMILE MCMLXIX

© WILLIAM T. STEARN AND D. S. FLETCHER

INTRODUCTION

BY

WILLIAM T. STEARN

William Curtis
1746–99
Botanist and Entomologist

MANY field naturalists have combined the study of entomology and botany. The close linkage between the lives of insects and plants being so evident to those who study them in nature, an interest in the one has often led to an interest in the other. Few naturalists, however, have maintained a life-long equal eminence in both these branches of natural history. One such was Adrian Hardy Haworth (1768–1833), the leading English authority on both succulent plants and British moths and butterflies in the early 19th century. When Haworth settled in 1792 at Chelsea, then a village in a rural district, he quickly became friendly with William Curtis (1746–99), who had been demonstrator of botany at the Chelsea Physic Garden from 1772 to 1777 and had lived since 1789 at 'Pond Place by the side of Chelsea Common' on the Fulham Road. Curtis's interests embraced entomology, as well as botany and gardening, and he undoubtedly encouraged young Haworth in its pursuit, to the extent indeed of giving

Haworth his collection of British insects. Haworth's reputation could as well rest upon his *Lepidoptera Britannica* (1803–28) as upon his *Observations on the Genus Mesembryanthemum* (1794–95) and *Synopsis Plantarum succulentarum* (1812). Curtis's monuments are botanical, his *Flora Londinensis* (1775–98) and *Curtis's Botanical Magazine* (1787 onwards) which he founded, and his place in British natural history would be insignificant but for these. Nevertheless Curtis's first publication was entomological, a little work entitled *Instructions for collecting and preserving Insects particularly Moths and Butterflies* (1771) and almost his last was likewise entomological, a posthumous paper 'On Aphides, the cause of honeydew on plants' in *Trans. Linnean Soc. London* 6: 75–94 (1802). His *Short History of the Brown-Tail Moth* appeared in 1782 when he was thirty-six. It is thus appropriate to accompany this facsimile of that now very scarce work, a pioneer essay in economic entomology, with some general notes on his career, of which detailed accounts were published by Thornton in 1803 and W. H. Curtis in 1941, and a bicentenary appreciation by J. E. Lousley in 1946.

William Curtis was born on 11 January 1746 at Alton, Hampshire, England, where his father John Curtis was a tanner. A strong medical tradition ran in the Curtis family, which was of Quaker persuasion, and the study of plants for materia medica naturally interested its members, as it did other Quakers: indeed George Fox himself had once intended to found an educational garden 'with all kind of physical plants for lads and lasses to learn simples there and the uses to convert them to'. His uncle James Curtis was a surgeon-apothecary at Alton; his grandfather John Curtis was also a surgeon-apothecary; and his great grandfather Thomas Curtis M.D. had been a physician. While a schoolboy his taste for natural history became evident. Given a small plot of ground in their Alton garden to cultivate as he pleased, he formed on this 'a wilderness of what to the vulgar eye appear weeds', not, however, from indolence but because they served as hosts to certain insects. His grandfather took him as an apprentice on leaving school, this being the usual way of entry into the medical profession then, particularly for dissenters, but as he seemed to devote too much of his time and attention to the natural history of the Alton countryside he was sent to London in 1766 to serve an apprenticeship with an apothecary, George Vaux. From him young Curtis moved to another London apothecary, Thomas Talwin, spending part of his time in his master's shop, part at lectures at St. Thomas' Hospital and part, too large a part for his master's taste, in botanical excursions. However, Talwin must have been sympathetic and Curtis genial, diligent and promising, for, when Talwin died, he bequeathed him his evidently prosperous business. Fortunately this continued to prosper and in 1770 Curtis took a partner by the name of Wavell, an arrangement which allowed him more time for his natural history studies. Later he sold the business to Wavell, being resolved henceforth to devote himself entirely to natural history. He was now twenty-five. In 1773 he established a garden at Bermondsey, south London, for the cultivation of native British plants, and began to give lectures on botany. Here he became friendly with an artist, William Kilburn, and engaged him to make coloured drawings of plants. In Denmark, J. G. Oeder had begun publication in 1761 of the *Flora Danica* with engraved and hand-coloured folio illustrations of Scandinavian plants; in Austria, N. J. Jacquin's *Flora Austriaca* began, in 1773, similarly to figure Central European plants. Earlier, between 1755 and 1760, Philip Miller had published in parts his *Figures of the most beautiful . . . Plants described in the Gardener's Dictionary* providing folio illustrations of cultivated exotic plants. Such works gave Curtis precedent and encouragement for the publication of a folio work illustrating British plants life-size, his superb *Flora Londinensis*. The first part of this, with six coloured plates, probably all of them drawn and engraved by Kilburn, appeared in May 1775. The previous year there had been published anonymously *A Catalogue of the Plants growing wild in the Environs of London*, which Dryander later attributed to Curtis; Gilmour (1949), however, has brought forward strong reasons for believing that Curtis had nothing whatever to do with this list compiled from Hudson's *Flora Anglica* (1762). In 1772 Curtis's reputation as a botanist led to his appointment as Praefectus Horti of the Chelsea Physic Garden, a post he held until 1777. During this period he published in the *Philosophical Transactions* of the Royal Society a list of 50 plants cultivated in the Chelsea Physic Garden, among them the new species *Saxifraga stolonifera* Curtis (1774), syn. *S. sarmentosa* Schreber (1780); cf. Fuchs, 1963. He also published another entomological work, *Fundamenta Entomologiae or an Introduction to the Knowledge of Insects* (1772), being a translation of Linnaeus's dissertation *Fundamenta Entomologiae* (1767), as printed in *Amoenitates Academicae* vol. 7 (1769), with some additions by Curtis.

The *Flora Londinensis* (1775–98) became Curtis's major achievement. This work consists of plates 'in folio, accurately drawn and neatly engraved', portraying native British plants 'drawn of natural size or, where a whole plant cannot be represented, a branch or part of it only', as stated by Curtis himself on the original wrappers. He intended to publish it in parts 'until the whole of the plants growing wild within ten miles of London are figured and described, and if the author should be favoured with health, and meet with the necessary encouragement, it is purposed afterwards to publish a continuation of it to extend to all the plants which are indigenous to Great Britain'. London as a built-up area did not then extend more than three miles west, one mile north and two miles east of St. Paul's Cathedral. It had not engulfed the villages of Battersea, Bow, Chelsea, Hackney, Islington, Paddington, etc. Bearded tits frequented the reeds along the Thames and snipe the undrained 'five fields' of Belgravia; as late as 1812 Montagu's harriers hunted the Battersea marshes. The diversity of habitats around London accordingly gave Curtis a great diversity of wild plants which could be illustrated from freshly gathered living material. Thus he wrote of *Dactylorhiza incarnata* (figured as *Orchis latifolia* in *Flora Londinensis* no. 59; 1787–88) that 'we need go no further than Battersea-Meadows to find this plant in tolerable abundance', together with buckbean (*Menyanthes trifoliata*, figured in no. 40; 1781–82), common cotton-grass (*Eriophorum angustifolium*, figured as *E. polystachion* in no. 37; 1781), marsh valerian (*Valeriana dioica*, figured in no. 47; 1783–85) and other marsh plants, which grew in the ditches and creeks by the Thames on the area which is now Battersea Park, its level having been raised, long after Curtis's death, by the dumping in the 19th century of many tons of soil brought there during the excavation of the docks lower down the Thames. He could refer to 'the cornfields around Peckham' and even 'Tothill-fields' at Westminster as the habitats of species which must now be sought many miles away. The now rare and vanishing marsh sow-thistle (*Sonchus palustris*, figured in no. 51; 1785) could then be found 'sparingly in the marshes around Blackwall and Poplar'. The snake's head fritillary (*Fritillaria meleagris*) grew 'in the meadows betwixt Mortlake and Kew'. Further down the Thames the neighbourhood of Charlton provided Curtis with many plants for figuring. The plates were drawn and engraved with exquisite skill and accuracy by William Kilburn, James Sowerby and Sydenham Edwards, with F. Sansom and W. Darton as additional engravers. Curtis supervised their work and wrote the text. This gives descriptions in both English and Latin, followed by a general account of the plant's habitat, occurrence and features of interest, including pharmacological and economic information when available and occasional quotations from Shakespeare, Milton and Dryden. One of the several Latin versions of the old riddle of the rose (cf. Stearn, 1965–6) probably first appeared in print in Curtis's account of *Rosa dumalis* (figured as *Rosa canina* in no. 51; 1785). Haworth quoted this version in his *Miscellanea naturalia* (1803) without mentioning the *Flora Londinensis* and he may well have first learned it direct from Curtis himself, since the riddle has obviously been handed down through the centuries by word of mouth.

Curtis hoped that his costly and ambitious undertaking would have a wide appeal and be well supported, particularly by the owners of 'the numerous pleasant villas, and delightful rural retreats of the nobility, gentry and opulent tradesmen, within the circle of ten miles from this metropolis, the abodes of ease and contemplation'. Alas, the general public, 'the student in botany . . . the botanist at a distance from London . . . the medical student . . . the farmer and grazier . . . the philosopher whose abstruse studies and sedentary life too often prove injurious to his health . . . those gentlemen and ladies, who from motives of profit and amusement occasionally employ themselves in painting flowers', all these potential supporters responded too little and publication of the *Flora Londinensis* was not a commercial success. Its production gave Curtis continual worry and distress. Subscribers looked forward so eagerly to new parts, each containing six folio plates, that they complained about the slowness of their appearance; some were dissatisfied with the colouring of their copies. The costs of drawing, engraving and printing outran Curtis's modest means; they swallowed up the money devised to him by his father's death, the profits of his lectures and the subscriptions to his garden, as stated by Thornton, and it was fortunate indeed that the philanthropic Dr. John Cloakley Lettsom, another Quaker, was wealthy enough to give him £500 in 1778 towards the continuance of the *Flora*.

A serious shortcoming was the lack of a consistent single system of numbering the plates. Some plates have numbers, most of them have none and hence had to have numbers added later by the owner; they were assigned one set of numbers in the index to the six fascicles and another set of numbers in the index to each volume. This made difficult both consultation and citation of the plates. Moreover the contents of the parts and their dates of publication were not stated in the work itself. Since Curtis published a number of new botanical names herein, i.e. *Ranunculus hirsutus*, *Bromus hirsutus*, *Typha major*, *Typha minor*, *Galeobdolon galeopsis*, *Carex riparia*, *Carex gracilis*, *Sisymbrium terrestre*, *Tormentilla officinalis*, *Chenopodium olidum*, *Geranium parviflorum*, *Carex ventricosa*, *Carduus tenuiflorus*, *Ophrys fucifera*, *Cerastium pumilum*, *Bromus diandrus*, *Galeopsis versicolor*, *Cerastium tetrandrum*, *Agrostis setacea* and *Poa procumbens*, some of which conflict with names published by other authors, this has necessitated much time-consuming bibliographical inquiry from 1881 onwards. Study of contemporary reviews and of a few parts in original wrappers as issued made it possible to list the contents of all the parts in the *Catalogue of Botanical Books in the Collection of Rachel M. M. Hunt* 2, ii: 392–409 (1961), but the precise dates of publication of most of the parts are still unknown; search for advertisements in contemporary London newspapers might reveal them. Most of the available information has come from contemporary notes and correspondence. Thus a receipt for no. 12 is dated 'April 22, 77', thereby indicating 1777 as the year of publication; according to other evidence it was issued not later than July 1777. On 4 May 1792 Goodenough wrote to Curtis: 'I am glad to find this auspicious day is to give birth to another daughter of *Flora*. I am so impatient for it, that I wish to know if you shall send anyone this way with your numbers'. This presumably refers to no. 62, since no. 61 contains plates dated 'Dec. 1, 1791' and, according to other evidence, no. 62 had certainly been published by 10 April 1793. In this way, little by little, approximate if not precise dates may be established. The first part appeared in May 1775, the 72nd and last part in 1798. The 432 engravings in this monumental work provide the finest series of coloured illustrations of British plants that have ever been published. They include not only flowering-plants and ferns but also some fungi and mosses. Modern identifications of the plates by J. E. Dandy were published in 1961 (cf. Stevenson, 1961).

In 1779 Curtis moved from Bermondsey and established the London Botanic Garden at Lambeth Marsh on the south bank of the Thames. Unfortunately the heavy smoke from London on the opposite bank of the Thames constantly enveloped his plants; when the time came for renewal of his lease in 1789, he moved them to another garden at Brompton, westward of the city of London and thus more conveniently placed for subscribers to the garden; here at last his plants grew 'in perfect health and vigour'.

Alarm in 1782 over the damage caused by the caterpillars of the Brown-tail moth 'at present uncommonly numerous and destructive in the vicinity of the Metropolis', when their large white webs were 'conspicuous on almost every hedge, tree and shrub' in the London area, led Curtis to publish then his *Short History* which remains of interest as much for its general methodology, for its effective combination of learning, firsthand knowledge, common-sense and logic, as for its particular subject, a long-forgotten insect epidemic. There existed then no professional economic entomologists, no plant pathologists, no government horticultural advisers backed by research institutes, to deal with such a matter and allay the public's fears.

Curtis therefore acted on his own 'to give the public a true idea of the nature of these Insects and thereby dispel their imaginary terrors'. A trained modern investigator would survey the affected area, search the literature for the known history of the pest, ascertain its range of hosts, elucidate its life-history, consider the effect of parasites and recommend control measures in an officially sponsored bulletin. He would have behind him the experience of generations of entomologists. Curtis lacked such backing but on his own initiative he tackled the matter in the same way; his *Short History*, the forerunner of innumerable reports on pests by government bodies, particularly in the United States, was, however, a private publication. In due course the numbers of the Brown-tail moth declined and Curtis's little work ceased to have any public interest in England. In 1897, however, the story began to be repeated in Massachusetts, to which the Brown-tail moth had been introduced on nursery stock from Europe and where it rose to be again an important pest of fruit and other deciduous trees. Curtis's *Short History*, although a tract for the times, clearly demonstrates both his quality as a naturalist and his readiness to put his extensive knowledge to public service.

By now Curtis had come to realize that the book-buying public was much more interested in exotic garden plants than in British weeds and other wild plants and that the folio format and the elaborate text of the *Flora Londinensis* made it too expensive for its sale adequately to cover the cost of production. A new undertaking ended his difficulties. In his garden he possessed a wide range of ornamental and interesting plants, many of which had never yet been illustrated. He discussed with close friends the publication of an octavo magazine illustrating such plants and giving their correct names, habitats and cultural requirements. He received every encouragement and the first number of *The Botanical Magazine; or, Flower-Garden displayed: in which the most ornamental foreign Plants, cultivated in the open Ground, the Green-house and the Stove will be accurately represented in their natural Colours* appeared in February 1787. It illustrated *Iris persica*, *Rudbeckia purpurea* (now *Echinacea purpurea*), *Helleborus hyemalis* (now *Eranthis hyemalis*); down to 1801 (t. 512) each part contained only three plates. Fortunately this new periodical proved an immediate success. It was attractive, convenient to handle and relatively cheap (one shilling a copy); it appeared regularly once a month; the plates were numbered in one sequence and could be readily cited. According to W. Hugh Curtis (1941), during Curtis's lifetime the average monthly sale is said to have been about two thousand copies; later the sales of the *Botanical Magazine* greatly declined, so that paradoxically the earlier numbers of the *Magazine* are much commoner than later ones, but, despite the competition of other periodicals with coloured plates, notably Andrews, *The Botanist's Repository* (1797–1812), Edwards and others, *The Botanical Register* (1815–47), Loddiges, *The Botanical Cabinet* (1818–33), Sweet, *The British Flower Garden* (1823–38) and Maund, *The Botanic Garden* (1825–51), the *Botanical Magazine* outlasted them all. In 1826 William Jackson Hooker became editor and gave the *Magazine* a more scientific character which it has never lost.

Hooker moved to Kew in 1841 and the *Magazine* has been intimately though unofficially associated with the Royal Botanic Gardens, Kew, ever since, being edited in turn by his three successors as director of the gardens, i.e. Joseph Dalton Hooker, William T. Thiselton-Dyer and David Prain. By the end of 1920, when its publishers Messrs. Lovell Reeve decided that they could no longer produce it at a loss, the *Magazine* had reached its 146th volume and had published 8,873 plates. Responsibility for its publication was then taken over by the Royal Horticultural Society of London and, when hand-colouring of the plates ceased in 1948, it had reached its 156th volume and published 9,688 plates. The provision of so many coloured illustrations of plants, with authoritative text, has been a service of inestimable value to systematic botany and to horticulture. Its initiation by Curtis must therefore rank as his most lasting contribution to both.

The sales of the *Botanical Magazine* made good the losses sustained over the *Flora Londinensis* and enabled Curtis and his family to live free of financial worry for the last ten years of his life, the last one of which was clouded by ill-health. He died on 7 July 1799, aged 53 years, and was buried at his own request in Battersea Churchyard because the neighbourhood had been one of 'frequent scenes of enjoyment in herbarizing excursions with his pupils', but the Thames-side marsh, with its orchids, buckbean and other exquisite wild flowers, and the weedy fields, which also made Battersea such a pleasant place for him, have vanished so completely that his *Flora Londinensis* alone stands as their memorial. Although it is true, as his friend Samuel Goodenough said in an obituary notice, that 'Mr. Curtis was perpetually forming some new design without completing any one', and that in the field of entomology he committed all too little to paper of his numerous observations, nevertheless his geniality, his learning and his readiness to help his many friends and students and the high standards he maintained in all he did, contributed much to entomology, botany and horticulture in his day. 'There was never a pleasanter companion than Mr. Curtis', according to Goodenough: 'he abounded in innocent mirth, and good humour ever floating uppermost, gave a pleasant cast to everything he said.' It is fitting that he bequeathed to posterity works which still convey to others the joy and the interest that living creatures were to him.

[11]
SOURCES OF FURTHER INFORMATION

ARDAGH, J. 1945. Curtis's botanic gardens. *Chronica Bot.* 9: 75, t. 9.

CHITTENDEN, F. J. 1956. *Curtis's Botanical Magazine Index . . . Volume 1 to Volume 164; to which is appended a brief History of the Magazine.* London (Royal Horticultural Society).

CURTIS, W. 1803. *Lectures on Botany as delivered to his Pupils, arranged by Samuel Curtis*, Vol. 1. London.

CURTIS, W. H. 1941. *William Curtis, 1746–1799, Fellow of the Linnean Society, Botanist and Entomologist.* Winchester (Warren & Son).

DARLINGTON, I. and HOWGEGO, J. 1964. *Printed Maps of London circa 1553–1850.* London (G. Philip & Son).

DREWITT, F. D. 1928. *The Romance of the Apothecaries' Garden at Chelsea.* 3rd ed. Cambridge.

FITTER, F. R. S. 1945. *London's Natural History.* London (Collins).

FUCHS, H. P. 1963. Nomenklatorische Erganzungen zu den Arbeit 'Nicolaas Meerburg und die drei von ihm verfassten botanischen Tafelwerke'. *Acta Bot. Neerl.* 12: 12–16.

GILMOUR, J. S. L. 1949. A 'Catalogue of London Plants' attributed to William Curtis. *J. Soc. Bibl. Nat. Hist.* 2: 181–182.

KEWENSIS [Goodenough, S.] 1799. William Curtis. *Gentleman's Mag.* 69: 635–639.

KITCHIN, T. 1773. *A pocket Plan of the Cities of London and Westminster.* London.

LOUSLEY, J. E. 1946. William Curtis (1746–1799). *J. R. Hort. Soc.* 71: 98–100, 124–129; *London Naturalist* 25: 3–12.

RAISTRICK, A. 1950. *Quakers in Science and Industry, being an Account of the Quaker Contributions to Science and Industry during the 17th and 18th Centuries.* London (Bannisdale Press).

STEARN, W. T. 1961. Botanical gardens and botanical literature in the eighteenth century. *Cat. Bot. Books Coll. R. M. M. Hunt* 2.ii: xlii–cxl. Pittsburgh (Hunt Botanical Library).

STEARN, W. T. 1965a. *Adrian Hardy Haworth, 1768–1833, Botanist, Entomologist and Gardener, his Life and Publications.* (Prefixed to A. H. Haworth, *Complete Works on succulent Plants*, Vol. 1, Gregg Press facsimile).

STEARN, W. T. 1965b. The five brethren of the rose; an old botanical riddle. *Huntia* 2: 180–184.

STEVENSON, A. 1961. *Catalogue of botanical Books in the Collection of Rachel McMasters Miller Hunt.* Vol. 2.ii: 389–412 [separately issued as 'A bibliographic study of William Curtis' Flora Londinensis' by A. Stevenson, J. E. Dandy and W. T. Stearn]. Pittsburgh (Hunt Botanical Library).

THORNTON, R. J. 1803. *Sketch of the Life and Writings of the late Mr. William Curtis.* London (prefixed to Vol. 1 of W. Curtis, *Lectures on Botany*).

ENTOMOLOGICAL NOTES

BY

D. S. FLETCHER

Dept. of Entomology, British Museum (Natural History)

THE BROWN-TAIL MOTH (*Euproctis chrysorrhoea* Linnaeus, known in the United States and Canada as *Nygmia phaeorrhoea* Donovan) belongs to a small, but destructive family of the moths, the Lymantriidae or Tussock moths.

Adult Lymantriidae are characterized by the absence of ocelli, by the proboscis being rudimentary or absent, and by the presence of tympanal organs in the thorax; in the male the antennae are bipectinate to the tip of the shaft. In the genus *Penthophera* the wings of the female are reduced, while in *Hemerocampa* and *Orgyia* the females are apterous. Females of the genus *Euproctis* have dense anal tufts of deciduous hair scales which, together with a glutinous secretion, are deposited over the egg masses.

Caterpillars are hairy and often bear dense lateral and dorsal tufts of hair scales (*Dasychira, Euproctis, Orgyia*), which give rise to the popular name for the family, Tussock Moths. The caterpillars of many species are gregarious and live in closely woven webs or bag-like nests during early instars (*Euproctis*), often occurring in vast numbers. The hairs of many species of *Euproctis* larvae, including those of *chrysorrhoea*, can cause a painful rash on contact with the human skin, though some people are more susceptible than others. The hairs are barbed and hollow and while still attached to the body of the larvae are connected to subcutaneous poison glands; even when shed, the dried poison within the hair may remain toxic for up to four years. Affected parts of the human body swell and the accompanying intense irritation may last for several hours. Eye injury is also possible through the penetration of the cornea of the eye by the cast larval hairs.

Pupation takes place above ground within a silken cocoon into which are woven larval hairs.

The Lymantriidae include several important pests of forest and orchard trees. The Gypsy moth (*Lymantria dispar* Linnaeus) of Europe and northern Asia, introduced accidentally into North America in 1868, is a serious forest and orchard pest. In India *Euproctis fraterna* Moore and *Euproctis lunata* Walker defoliate castor, *Euproctis howra* Moore defoliates coffee and *Euproctis scintillans* Walker attacks a variety of crops. In South Africa *Euproctis terminalis* Walker, in addition to feeding on the foliage of several shade trees, has become a serious pest of introduced pines.

Euproctis chrysorrhoea Linnaeus is a pest of forest trees, orchards and hedgerows in much of Europe and in the New England States of North America, where it was introduced accidentally at the end of last century. In Britain the Brown-tail moth is confined to S.E. England, occurring mainly in the coastal regions of Suffolk, Essex, Kent, Sussex and Hampshire, including the Isle of Wight. Elsewhere in Europe the species extends from Southern Scandinavia to Algeria and Tunisia and eastwards to the Western Urals, Transcaucasia and Syria.

Periodically populations build up to epidemic proportions and completely defoliate forest and shade trees, orchards and hedgerows. Such outbreaks, similar to that in London described by Curtis in the latter half of the 18th century, occurred in Essex in 1947–48, in European Russia in 1949–55, in Austria and Germany in 1952–56, in Northern Italy in 1955 and in Yugoslavia in 1957–59.

Control is effected by the natural increase of dipterous and hymenopterous parasites, by the collection and destruction of larval nests in winter and by the use of chemical and biological dusts and sprays. Applications of either DDT insecticide or pure culture of *Bacillus thuringiensis* var. *dendrolimus* (dendrobacillin) or a mixture of both to larvae feeding after hibernation, have proved highly effective in the Ukraine. Cultures of *Bacillus cereus* var. *galleriae* (entobacterin) used in the U.S.S.R. had its greatest effect on fourth instar larvae and possessed the advantage of leaving natural parasites unaffected. Control by food factor has also been noted. The secondary growth of trees defoliated by spring-feeding larvae has been found to be deficient in cellulose and sugar; the following generation of autumn-feeding larvae grow rapidly, but are poorly adapted for hibernation and ensuing mortality is high. When populations are high and overcrowded conditions exist for the larvae, outbreaks of fungus diseases, particularly of *Entomophthora aulicae* Reiche, occasionally become epidemic and result in their almost complete extermination.

The Brown-tail was first discovered in North America in 1897 at Somerville, Massachusetts, where it was believed to have been introduced some years earlier with imported nursery stock. Vigorous control measures were taken until early in 1900 when numbers of the insect were so reduced that these measures were discontinued. As a result, numbers increased to pest proportions in Massachusetts, Rhode Island, New Hampshire and Maine, and continued to increase until 1915. During this period large areas of oak woodland in addition to shade trees and fruit trees were completely defoliated.

In 1905 the Brown-tail spread to Nova Scotia, where it soon became a serious pest to fruit growers. However, systematic collection and burning of winter nests, combined with spraying programmes, effected almost complete control. No record of the species exists in the province for the decade 1931–40 and there are only isolated records for the period 1941–50.

In 1906 a number of dipterous and hymenopterous parasites of the Brown-tail moth were introduced into the United States from Europe; the Tachinids *Sturmia nidicola* Townsend, *Carcelia laxifrons* Villeneuve and the Braconids, *Apanteles lacteicolor* Viereck and *Meteorus versicolor* Wesmael proved the most successful. Another Tachinid, *Compsilura concinnata* Meigen, successfully introduced from Europe, is parasitic both on the Brown-tail and Gypsy moths. Native parasites have had little effect on the Brown-tail moth in North America, but populations of larvae and pupae have been almost completely exterminated in overcrowded conditions by the fungus disease *Entomophthora aulicae* Reiche. Extremely low winter temperatures have also restricted the spread of the Brown-tail; temperatures of $-25°F$. and below have caused high mortality among larvae in unsheltered nests or in nests unprotected by snow. Biological control has been reinforced with the use of chemical sprays and by the cutting and burning of larval nests in winter. During the winter of 1933–34 a total of nearly 24 million nests were collected and destroyed, in the New England States.

Strict quarantine regulations control the movement of forest products and nursery stock in a barrier zone surrounding the areas of New England infested by the Brown-tail moth; no article in these categories may be moved without a Federal certificate of inspection. By a combination of control measures the pest has been successfully confined to the New England States in the U.S.A. and to New Brunswick and Nova Scotia in Canada.

REFERENCES

Androic, M. 1963. Factors which impede the successful control of *Euproctis chrysorrhoea*. *Plant Prot.* 14: 273–285, 5 figs.

Anon. 1956. The Brown-tail Moth, how we fight it. *U.S. Dept. Agric.*, PA. 282.

Burgess, A. F. & Baker, W. L. 1938. The Gypsy and Brown-tail moths and their control. *U.S. Dept. Agric., Circ. No.* 464. 38 pp., 17 figs. Washington.

Burgess, A. F. & Crossman, S. S. 1929. Imported Insect Enemies of the Gypsy moth and the Brown-tail moth. *U.S. Dept. Agric., Tech. Bull. No.* 86. 148 pp., 6 pls., 55 text figs., 34 tables.

Grobler, J. H. 1957. Some Aspects of the Biology, Ecology and Control of the Pine Brown Tail Moth, *Euproctis terminalis* Walker. 186 pp., 142 figs., 37 tables.

Kam'yanyi, L. A. 1964. Tests of small-scale aerial spraying for the control of forest pests. *Zakhyst Roslyn*, pt. 1: 84–88.

Lappa, N. V. 1963. On the susceptibility of larvae of the brown-tail moth to infection with *Bacillus cereus* var. *galleriae*. *Zool. Zh.* 42: 1064–70, 1 fig.

——1964. The action of the bacterial preparation dendrobacillin on larvae of the brown-tail moth. *Zakhyst Roslyn*, pt. 1: 65–72.

Naumov, R. V. 1959. On the causes of the cessation of mass outbreaks of *Euproctis chrysorrhoea*. *Zool. Zh.* 38: 133–134.

Pantyukhov, G. A. 1964. The effect of negative temperatures on populations of the brown-tail moth *Euproctis chrysorrhoea* L. and the gypsy moth *Lymantria dispar* L. (Lepidoptera Orgyidae). *Ent. Obozr.* 43: 94–111, 13 figs.

Saccuman, G. 1963. Contributo alla conoscenza della *Euproctis chrysorrhoea* L. Cenni sulla morfologia, biologia e mezzi di lotta. *Boll. Lab. Ent. agric. Portici*, 21: 271–322, 12 figs.

SMIT, B. 1964. Insects in Southern Africa, How to Control them. 399 pp., 228 figs. Cape Town.

TEMPLIN, E. 1958. Der Einfluss von Bekämpfungsaktionen auf den Verlauf der letzten Gradation von *Euproctis chrysorrhoea* L. *Z. angew. Ent.* 41: 425–437, 3 maps.

VANKOVA, J. 1957. Study of the Effect of *Bacillus thuringiensis* on Insects. *Folia Biol.* 3: 175–182, 1 pl.

WATSON, P. G. & SEVEL, D. 1966. Ophthalmia nodosa. *Brit. J. ophthal.* 50: 209–217, 7 figs.

A SHORT

HISTORY

OF THE

BROWN-TAIL MOTH,

THE

CATERPILLARS

of which are at prefent uncommonly numerous and
deftructive in the Vicinity of the Metropolis.

Illuftrated by a COPPER-PLATE, coloured from Nature,
reprefenting the Infect in its various States.

By WILLIAM CURTIS,
AUTHOR OF THE FLORA LONDINENSIS.

LONDON:
Publifhed by B. WHITE, Fleet-ftreet; J. SEWELL, Cornhill; J. JOHNSON,
St. Paul's Church-yard; J. STRAHAN, N° 67, Strand; and R. FAULDER,
New Bond-ftreet.
MDCCLXXXII.

A SHORT HISTORY OF THE BROWN-TAIL MOTH, &c.

THE attention of the public has of late been strongly excited by the unusual appearance of infinite * numbers of large white webs, containing caterpillars, conspicuous on almost every hedge, tree, and shrub, in the vicinity of the metropolis; respecting which, advertisements, paragraphs, letters, &c. almost without number, have appeared in the several news-papers, most of which, though written with a good design, have tended greatly to alarm the minds of the people,

---

* Some idea may be formed of their numbers from the following circumstance. In many of the parishes about London, subscriptions have been opened, and the poor people employed to cut off and collect the webs at one shilling *per* bushel, which have been burned under the inspection of the church-wardens, overseers, or beadle, of the parish: at the first onset of this business, four-score bushels, as I was most credibly informed, were collected in one day in the parish of Clapham.

A 2   especially

especially the weak and the timid. Some of those writers have gone so far as to assert, that they were an usual presage of the plague; others, that their numbers were great enough to render the air pestilential, and that they would mangle and destroy every kind of vegetable, and starve the cattle in the fields. From these alarming misrepresentations almost every one, ignorant of their history, has been under some dismal apprehensions concerning them; and even prayers have been offered up in some churches, to deliver us from the apprehended approaching calamity.

To give the public a true idea of the nature of these Insects, and thereby dispel their imaginary terrors; to shew what the mischiefs are which they are really capable of occasioning, and to point out the most likely means of obviating those mischiefs, are the motives which induce me to collect together and publish the notes and observations I have from time to time made concerning them, not as containing the compleatest possible history of the Insect, but such as may be expected on the spur of the occasion.

It may be remarked, in the first place, that the Insect in question is not new in this country, being every year to be found in abundance, and well known to those who collect Insects to be the Caterpillar of the *Brown-tail Moth:* nor is it peculiar to this country, but found in many parts of Europe, and has been considered, by all who have written on it, as notorious for its ravages. ALBIN, an English writer on Insects, 1720, says, that the Caterpillars of this moth lay themselves up in webs all winter, and as soon as the *Buds* open, they come forth and devour them in such a manner, that whole trees, and sometimes hedges, for a great way together, are absolutely bare. GEOFFROY, a French author, in his History of the Insects about Paris, describes it as the most common of any with them; that it is found on most of their trees, which it often strips entirely of their foliage in the spring *.

Our great naturalist RAY also describes this Caterpillar in his *Historia Insectorum* †. It is likewise figured and described by ROESEL, a German writer. LINNÆUS has either omitted, or confounded it with the *Phalæna Chrysorrhæa*, or *Yellow-tail Moth*, with which it has a great affinity.

These authorities will be sufficient to shew, that it is no new Insect, and that its ravages are not unusual. It must, however, be allowed that they are, and have been the two last years, uncommonly great, and unusually extensive.

When Insects are multiplied in this extraordinary manner, it is seldom that they extend through a whole country: the precise tract which these occupy I have had no opportunity of observing. On the Kingston Road I traced them as far as Putney Common, on the farther part of which, on the trees about Coomb Wood and Richmond Park, a web was not to be seen. I remarked, that they were extremely numerous to the distance of about eight miles on the Uxbridge Road. On the Great Western Road they terminated about the Star and Garter leading to Kew; from whence to Alton in Hampshire, not one was visible; and I have received undoubted information from other quarters, that the destruction they occasion is by no means general.

* Sa Chenille a seize pattes. C'est la plus commune de toutes. Elle est velue, de couleur jaunâtre, et elle vient sur presque tous les arbres, qu'elle dépouille souvent entiérement des les printems. GEOFFROY *Hist. abrégee des Insectes qui se trouvent aux Environs de Paris*, p. 117.

† Thus, *Eruca longis pilis fulvis hirsuta pulla, punctis albis et duobus rubris in imo dorso varia*, p. 347.

Having

Having shewn, that this insect is neither new in its appearance, nor general in its ravages, I shall endeavour to demonstrate, that there is no reason to be so dreadfully alarmed about its effects, as its powers of destruction are much more limited than is generally imagined.

Experiment teaches us, that there are some Caterpillars which are designed to feed on one kind of plant only, as the *Papilio Urticæ*, and *Iö, small Tortoise-shell* and *Peacock Butterflies*; these are never found on any other plant than the stinging nettle. Others that are attached to two or more sorts, as the *Phalæna Verbasci*, or *Water Betony Moth*, which appears to be equally fond of the *Mullein* and *Water Betony*: while others will devour indiscriminately almost every kind of herb, shrub, or tree, as the *Phalæna Antiqua*, or *Vapourer Moth*, which I have seen to thrive on the *deadly Nightshade* and *poisonous Laurel*.

The present Caterpillar is not so limited a feeder as some, nor so general a one as others. Its whole œconomy, however, shews it designed to feed on trees and shrubs, on which alone it is ever found. These afford it a support for its web, which is an habitation in many respects essential to its existence, and with which herbaceous plants cannot supply it.

We may, therefore, with as much propriety expect to see the *Cabbage Butterfly*, *Papilio Brassicæ*, feed on our Oaks and Elms, as to find these Caterpillars destroying our Herbs or Grass.

The following facts will serve to corroborate what is here advanced. They are found on the

Hawthorn most plentifully.
Oak the same.
Elm very plentifully.
Most *fruit trees* the same.
Blackthorn

Blackthorn plentifully.
Rose trees the same.
Bramble the same.
On the *Willow* and
Poplar scarce.
None have been noticed on the
Elder.
Walnut.
Ash.
Fir, or
Herbaceous Plants.

Thus it appears, that the only mischiefs these Caterpillars are capable of occasioning, is to rob particular trees and shrubs of their foliage and blossoms: it remains to consider how far the trees and shrubs will be injured by such a loss? and how far it may be injurious to their owners? I have found, by repeated observation, that those trees and shrubs which have been entirely stripped have not been killed thereby, but as soon as the Caterpillars have removed to change to Chrysalis, they have put forth fresh foliage: the only loss, therefore, the owner sustains from their depredations on those trees which are not cultivated for the sake of their fruit, is some check to their growth, and a temporary deprivation of the beauties of spring and autumn. With respect to fruit trees, the injuries they sustain are more serious ones; as in destroying the blossoms, as yet in the bud, they also destroy the fruit in embryo: the owners of orchards and standard fruit trees have therefore most reason to be alarmed.

The idea of their producing the plague, &c. is founded in the grossest ignorance, and carries with it its own refutation;

the

the health of the public is not, nor cannot be affected by them, either immediately or remotely.

Some persons have been alarmed leaſt, as they have now increaſed for three ſucceſſive years, they ſhould be infinitely more numerous the next. It may afford ſome ſatisfaction to thoſe to be informed, that it is extremely probable, the trees and ſhrubs will not afford ſufficient ſuſtenance to the preſent accumulated brood; for ſhould they be in the leaſt ſtinted in their growth at the time of their changing to Chryſalis, their wings will never expand, they will be incapable of flying, and of propagating their ſpecies. It is alſo extremely improbable, that the ſame circumſtances ſhould favour their increaſe another year.

What the actual cauſes are which occaſionally produce theſe extraordinary quantities of inſects will, perhaps, for ever remain among the arcana of nature. We frequently hear that, in certain parts of the country, much damage has been ſuſtained by the *Cock Chafer (Scarabæus Melolontha)*; in others, that the turnips have been deſtroyed by a ſmall Beetle, called the *Fly*; in others, that the ſame plant has been conſumed by a Caterpillar of a *Tenthredo* *; in others, that the wheat has been eaten in the ground by a ſmall grub, producing an *Elater*, or *Spring Beetle* †. One year the Aurelian ſhall find plenty of *painted Ladies (Papilio Cardui)*; another year, with all his care, he ſhall not find a ſingle one. Laſt year the *Sphinx Convolvuli*, *Convolvulus Hawk Moth*, and *Papilio Hyale*, clouded

---

* For an account of which ſee ALBIN, plate 62. Theſe, Mr. FRANKLAND informed me, were highly injurious laſt year in ſome parts of Yorkſhire.

† Mr. LIGHTFOOT ſhewed me ſome of theſe grubs; and related to me, that they were ſo deſtructive this ſpring about Uxbridge, that many farmers would be obliged to ſow freſh grain.

Yellow

Yellow Butterfly, were common about London; the naturalist may, perhaps, wait six years before he sees another.

The most probable causes are, the peculiarity of the weather, and the plenty or scarcity of the enemies of the Insect; for almost every different species of Insect has its peculiar enemy, and none more than the Caterpillars of Moths and Butterflies. As to the former, warm and dry weather are universally allowed to promote the generation of Insects; violent winds, heavy and long continued rains, or extreme cold, are, on the contrary, supposed to check and destroy them. It is, however, wonderful to observe, with what address they secure themselves from the effects of the two former; such as feed on the boughs, on such occasions creep from them to the large branches or body of the tree, where they rest unshaken; and those who reside in webs are so secured as to suffer little injury from any of those causes.

We observed, that Caterpillars had many enemies. Birds of various kinds feed on them: the stomach of a cuckow that was shot, was found full of the Caterpillars of the *Buff-tip Moth (Ph. Bucephala)*. Mr. CHURCH, Surgeon, of Islington, has observed birds very busily feeding on the Caterpillars of this very Moth, and carrying them to their young. The Earwig is a great destroyer of Caterpillars. But their grand enemy is the Ichneumon Fly*, a proof of whose destructive powers I experienced

---

* There are various kinds of flies which pierce the skin, and deposit their eggs in the bodies of Caterpillars; but the most common is the *Ichneumon Fly*. The eggs thus laid quickly produce small maggots, which feed on the body of the Caterpillar, taking care to eat that part only which lies immediately under the skin, whereby they avoid injuring those parts which are essential to life; for, should the Caterpillar be destroyed, they also would perish. Till the period

experienced last year, in attempting to breed the *large garden white Butterfly, Papilio Brassicæ*. Out of twenty Caterpillars taken from the Cabbage, eighteen were stung by this Insect, and died. In proportion then as the Insect's enemies are more or less numerous, so may be the Insect itself.

We shall now proceed to give a short account of the history of the Insect which is the subject of this essay; in the course of which, we shall point out what appears to us to be the best and most effectual method of destroying them. The Caterpillar, as already has been observed, owes its origin to a Moth, called the *Brown-tail Moth*, which is about two-thirds of the size of the Moth produced from the Silk-worm, and is of a white colour throughout, excepting a streak of brown on the under side of each fore-wing, running near to, and parallel with, its anterior edge, and a brown or mouse-coloured tail, from whence it derives its name. These Moths come out of Chrysalis about the beginning of July, at which time they may be found flying about slowly, especially in the evening, and depositing their eggs on the foliage of the trees and shrubs before mentioned. The female has a much larger tuft of down on its tail than the male, a great part of which is made

of their full growth arrives, the Caterpillar eats as usual, and appears equally well in health. That period arrived, which seldom happens till the Caterpillar has crept to some convenient place to chrysalize in, they eat their way out of its body, and immediately spin themselves small bags, or cases, in which they change to Chrysalis. During this operation, which continues several days, the Caterpillar apparently suffers the greatest pain, and after struggling in vain with its mortal enemy dies. When the Ichneumon Fly is small, its maggots are proportionably numerous; a hundred of them frequently proceed from one Caterpillar. The little bags which they spin to chrysalize in are frequently mistaken by the ignorant for the eggs of the Caterpillar.

use

use of in covering its eggs, which, when laid, look like small lumps of down on the leaves.

The young Caterpillars are hatched early in autumn: as soon as they quit the egg they set about spinning a web, and having formed a small one, they proceed to feed on the foliage, by eating the upper surface and fleshy part of the leaf, and leaving the under side and ribs. It is curious to observe with what regularity they marshal themselves for this purpose. Thus they proceed daily, spinning and enlarging their web, to which they retreat every night and in bad weather, and extending their depredations. In the course of a few weeks their operations begin to be visible on the trees; their web as yet is not so conspicuous as those leaves, which, being stripped of their green part, assume a dead appearance: now is the time to destroy them, while their nest is small, and their ravages just conspicuous. They may be cut off the twigs or branches with a pruning knife, or gardener's shears, whose handles may, if necessary, be lengthened; or by a sharp hook affixed to the end of a long pole. When cut off, they should be collected together and burned, merely to prevent their returning again to the trees and shrubs. By performing this operation thus early, you save the autumnal verdure of your foliage: if it be deferred till winter, the web will then be more conspicuous, and will have acquired a stronger and tougher texture, so as to bear pulling off, which should be preferred to pruning in certain cases, especially where it regards fruit trees. No remedy short of removing the webs will avail. Lotions, fumigations, vermin powder, &c. will be applied to no purpose; they are too strongly enveloped to be affected by any of these. In about three weeks from their being first hatched, they change their skin, a process which not only all *Caterpillars* undergo four or five

five times, at different periods of their growth, but also the *Spider*, the *Bed Bug*, and even *Lobsters* and *Crabs*. This usually takes up several days. Afterwards, they proceed in the same manner, enlarging their web, and extending their daily foraging excursions, till benumbing winter confines them entirely to their silken habitation; they then not only secure the general web on all sides as strongly as they can, to exclude impertinent intruders, but each individual spins a thin case for itself: here they rest in a state of torpid security, till the genial warmth of the spring animates them afresh, and informs them, that the all-bountiful Author of Nature hath provided food convenient for them. Thus apprized, they issue forth in the day-time and in fine weather, as before; but having acquired stronger powers, and the foliage they have now to encounter being more tender, they become less scrupulous in their feeding, and devour the whole of it. A disposition to associate continues with them till they have changed their last skins, when they usually separate, each endeavouring to provide in the best manner for itself. At this period they are most exposed to various enemies, and most frequently attacked by the Ichneumon Fly (vide fig. 14, 15.). We sometimes find a few continuing together to the last, when each spins a separate web, in which it changes to Chrysalis: this usually takes place about the beginning of June; here, in a state of perfect quietude, it remains about three weeks, when it changes to the Moth we have already described.

EXPLANATION OF THE PLATE.

Fig. 1. The eggs laid by the female Moth, and covered by the down from its tail.

2. The eggs with the down removed.
3. The young Caterpillars suspending themselves by a single web from their mouths.
4. Shews the manner in which the young Caterpillars feed from the time they are hatched till winter, by eating the surface and fleshy part of the leaf, and leaving the membranous and veiny part.
5. The general bag or nest of one brood of Caterpillars as it appears in winter.
6. Some of the Caterpillars as they appear on their first coming out in the spring.
7. A full-grown Caterpillar.
8. The same, having spun a web, and about changing to Chrysalis.
9. The same, changed to Chrysalis, and taken from the web.
10. A male Moth, produced from the Chrysalis.
11. A female of the same species.
12. A dead Caterpillar of the same species, having been stung by an Ichneumon Fly.
13. The web which the Caterpillar had spun, as at fig. 8. opened to shew the little bags which the Ichneumon Caterpillar, proceeding from its body, had spun to chrysalize in.
14. The Ichneumon Fly of its natural size, produced from a Chrysalis inclosed in one of the said bags or cases.
15. The same magnified.